Rocky Mountain Wildlife

ROCKY MOUNTAIN WILDLIFE

Ron Parker

RAINCOAST BOOKS

Vancouver

First published in 1999 by

Raincoast Books
8680 Cambie Street
Vancouver, B.C.
v6p 6m9
(604) 323-7100

www.raincoast.com

1 2 3 4 5 6 7 8 9 10

Canadian Cataloguing in Publication Data
Parker, Ron.
  Rocky Mountain wildlife

ISBN 1-55192-191-X

1. Mammals—Rocky Mountains, Canadian (B.C. and Alta.)—Pictorial works.*
2. Birds—Rocky Mountains, Canadian (B.C. and Alta.)—Pictorial works.*
3. Animals in art. I. Title.
QL221.R6P37 1999   591.09711   C98-911154-7

THE CANADA COUNCIL | LE CONSEIL DES ARTS
FOR THE ARTS | DU CANADA
SINCE 1957 | DEPUIS 1957

Raincoast Books gratefully acknowledges the support of the Government of Canada, through the Book Publishing Industry Development Program, the Canada Council and the Department of Canadian Heritage. We also acknowledge the assistance of the Province of British Columbia, through the British Columbia Arts Council.

Printed in Canada.

PREFACE

Every season of the year in the Rocky Mountains holds its enchantments. Each yields a different yet equally compelling chance to observe wildlife in its natural state.

Spring is a time for renewal. New shoots are pushing through the brown earth and dead grasses. Bright green buds are exploding on barren tree branches, and the skies are filled with birds returning from their wintering sites in the south.

Ducks and geese pass through the Rocky Mountains, heading for their breeding grounds on Arctic shores, coastal estuaries, lakes and prairie potholes of the great plains. Resident birds like chickadees and bald eagles are busy building new nests and repairing old ones.

While spring may be the best time of year for bird watching, it is not the time to see animals in their prime. Winter coats are being shed, making the deer, mountain goats and wolves look rather shabby. One compensation is the babies. Young animals are born in the spring, but their parents hide newborns away while they are most vulnerable, so they are difficult to find and observe. Doing so is a privilege which, once enjoyed, is never forgotten.

Summer brings abundant growth in the alpine with the beautiful colors and heady aromas of carpets of wildflowers. The heat of the summer sun makes the rocky slopes shimmer like mirages. Viewed from high mountain passes, ranges of blue-tinted mountains disappear into the distance, the cerulean sky pierced by their white peaks. Activity abounds as the wolf cubs grow and play. The racks of elk and deer support a soft, velvety skin and their bright summer coats shine with healthy vigor.

Early summer is a great time of year for me. I can finally get out and take some long hikes and get up into the high country. However, it is late summer that draws me to the Rocky Mountains. The creeks are no longer swollen with torrents of melting snow. The high country is warm and dry, making hiking above the tree line easy.

The animals are in the high country, the mountain passes and the high plateaus. It is then that a lone hiker can watch and photograph bighorn sheep, elk, caribou, moose and grizzlies in areas where trees are sparse and you can see forever. Eagles soar above the spires of lofty peaks, and deer graze on the open meadows.

Autumn brings dramatic changes. The morning air is brisk and invigorating, and the autumn colors are splendid. The yellow of the aspen leaves and gold of the larches contrast with the dark greens of the mature conifers. In the high mountain passes, the willows turn a variety of yellows, and the birches many hues of red and burgundy. In the mornings, the icy streams have edges of clear, freshly formed ice making narrow shelves around the rocks, and the golden grasses crunch with frost.

Autumn is my favorite time of year. This is the time when hiking and animal viewing are at their best. In the mountain parks, the summer hikers are gone, and a backpacker can be alone in the wilderness … alone with the vast valleys and mountain ranges, alone with the animals.

From a perch high up on the side of a mountain pass, an observer can watch the caribou prance by with heads held high, frosty breath snorting from their nostrils. The peaks ring with the bugling calls of the elk, and grizzlies lumber by with their thick winter coats shimmering and flowing like the windy undulations of summer fields. The bucks have full sets of antlers bare of the summer velvet and shining like polished bone It is also the time of the rut, and elk and deer bulls alike have the thick neck and physical vitality that mark this season. Creeks are at their lowest, making them easy to ford, and even though nighttime temperatures are often below freezing, daytime temperatures can rise almost as high as in summer.

Along the Rocky Mountain flyway, birds are heading south by the hundreds. Honks from skeins of geese fill the air as they begin their migration south for the winter along with flocks of songbirds and swans.

Winter seems to me to be a time of silence. Gone is the bustling energy of autumn. The animals that are not hibernating are conserving their energies to be able to last the winter on accumulated fat and the meager gleanings they can find under the snow. Squirrels and chipmunks feed in the stores gathered in the autumn, buried and hidden. Ground squirrels are deep in hibernation, and bears are snug in their dens, sleeping through the winter.

Mountain goats stay on the windblown slopes, pawing away to get at the frozen grasses, while deer and elk do the same in the valleys. The few birds that remain fluff up their feathers to keep out the cold and feed on the seeds and insects still on the weeds and trees.

I love to get an early start on an icy morning. With felt packboots, longjohns, a down parka, mitts and a pair of snowshoes, I can last for several hours out in subzero weather. It is a wonderfully invigorating feeling to crunch through the snow, camera in hand, breathing in the icy air and breathing out clouds of vapor. One of the most enjoyable experiences I've ever had was to snowshoe out through a local alpine area before dawn, under the light of a full moon. It was wonderful to stump through the sparkling snow, absolutely alone in the frigid silence with the white moon shining above the trees in an indigo sky.

It is always a pleasure to see bald eagles flying above the water or through the towering trees of the Pacific coast, but it is an even greater thrill to see them in the mountains. In the dry, mountainous areas on the west side of the Rockies, bald eagles must share air space with golden eagles and osprey.

From a treetop perch, high on a cliffside, a pair of bald eagles surveys its domain. Little escapes the eagle's baleful glare. Opportunistic hunters, with a special taste for fish, eagles can be seen feeding on spawning Kokanee salmon in the mountain streams. An eagle will ride the upwelling thermals along a mountain ridge, circling over-head, then swoop down for the kill, grace and menace joined in one swift motion.

Bald Eagle   *Eagles in the Pine*
24" x 36" (61 cm x 91 cm)

In the soggy snows of late spring, a hungry cougar can easily follow the tracks of its prey. Although it may dine on smaller fare, the cougar will stalk and bring down deer and even elk. Especially in the lean months of winter and early spring, one big kill can sustain a cougar for days.

Cougars are solitary creatures, living and hunting in their own home territories. The biggest cats in the North America, male cougars can grow to nearly nine feet (three meters) from their whiskers to the tip of their tails and weigh up to 200 pounds (80 kilograms). Young cougar cubs have a spotted coat, but as they mature, their fur takes on its characteristic tawny color.

Cougar   *Soft Snow*
20" x 40" (50 cm x 100 cm)

Canada geese are harbingers of spring in the Rockies. As soon as there is some open water, they arrive, either as passers-by on their way north, or as residents ready to take up nest building on one of the many lakes or marshes. This pair has landed on a stream recently divested of its winter ice. They will seek out a safe nesting place on shore.

In a nest lined with moss and down, the female will lay from three to six eggs. Once the goslings hatch, both parents care for the young. These geese may just spend a few months in the Rockies, but, if they are lucky, they will be back next year. And chances are they will come together: geese mate for life.

Canada Geese   *Spring Arrivals*
20" x 36" (50 cm x 91 cm)

Venturing forth from the network of tunnels it shares with others of its kind, the mantled ground squirrel can be seen darting among the scree and boulders on the mountain slopes. Like its close relation the squirrel, it stuffs seeds, nuts and berries into its cheek pouches to be stashed away later. After a long winter's hibernation, the mantled ground squirrel revels in the warm spring sunshine which melts the snows, uncovering hidden treasures.

Ground Squirrel   *Mountain Blooms*
18" x 14" (45 cm x 35 cm)

This bold little red fox just had to satisfy its curiosity by popping out of the den to look at the large creature watching and photographing it. The vixen, its mother, stayed close by, dropping in and out of other, more hidden entrances to the den.

This pup has plenty to learn. By the time it is six months old, it is more or less on its own. It subsists on the foods offered by the forest floor: seeds, insects, small rodents and amphibians and the occasional bird's egg. It appears cute at this age, with its black-tipped ears and bushy tail. But do not be fooled: the fox is a crafty and efficient killer.

Red Fox *Fox Pup at Den Entrance*
24" x 18" (61 cm x 45 cm)

No sound is more suggestive of the northern wilderness than the haunting call of the loon. Its mournful wail is a call for company, while what seems like maniacal laughter is actually a warning of a threat at hand. Warm-weather residents of the Rocky Mountain lakes, loons can be seen gliding gracefully, then plunging down to snatch a fish. The loon is a swimmer, first and foremost. An ungainly walker, it seldom ventures onto shore. When it takes flight, it dances along the surface of the water some distance before gaining the air.

To get a loon's-eye view of the shoreline, I was up at dawn and waded chest deep out into the frigid lake to capture the early morning light on the rocks and shore grasses. It proved to be the perfect setting for *Summer Loon.*

Loon   *Summer Loon*
18" x 24" (45 cm x 61 cm)

Attracted by splashing sounds, I scrambled out of the trees and brush to find myself face-to-face with a fully grown moose, head emerging from the lake and water dripping off his magnificent antlers. This encounter inspired me to paint *Autumn Foraging.*

The moose, the largest member of the deer family, frequents the coniferous forests of the Rockies, but it has a special affinity for spongy marshlands, across which its broad hooves allow it to maneuver nimbly. In summer, the marshes offer abundant plant life to stoke a moose's massive body, which can weigh close to a ton. Moose range alone or in small groups throughout the warmer months. In winter, they congregate in larger herds. Is it for warmth or protection, or do they just seek company through the cold, dark, snowy time of year?

Moose   *Autumn Foraging*
24" x 40" (61 cm x 101 cm)

When we lived in the Kootenay River valley, one of my favorite day trips was into nearby Kootenay National Park to see the elk. We would leave just before dawn and arrive at the meadows as the sky lightened and the sun just grazed the tops of the Rockies. Herds of elk abounded in the valley, and the mountains resounded with the bugling of the bulls.

It was just such a scene that I have depicted in *Autumn Meadow.* This early in the day, the air is crisp and the warm breath of the six-point bull creates a small mist in front of him. When he bugles, he will lay his huge rack against his back as he lifts his head and blasts out a lengthy, high-pitched call. This bugling is used both to attract females to his herd and to warn other males of his presence.

Elk   *Autumn Meadow*
24" x 36" (61 CM x 91 CM)

A young black bear cub, here enjoying its first summer, will spend up to a year in the company of its mother. The grizzly bear has a fearsome reputation, but the black bear can be just as dangerous to humans, especially when a sow feels its cub is threatened.

In the lush summer season black bears forage on nutrient-rich succulents. They eat a balanced diet, mixing in berries and other vegetation with the flesh of fish and animals. They do not truly sleep right through the winter, but the fat they pack on through active feeding in the summer and fall allows them to spend most of their time sleeping in their dens when the wind-driven snow howls outside.

Black Bear   *Curious Cub*
18" x 24" (45 cm x 61 cm)

The late September sun has lost some of its strength, here, in the northern Rockies. The first snows are not melting off the mountains and the air remains crisp til noon. Wolves are on the move, seeking better hunting in another valley, taking advantage of this snow-free time, before the first real blizzards close access. Looking into the eyes of a wolf facing me in the mountains is one of my most thrilling experiences. The eyes are intelligent and curious. They cast an aura of independence that makes you realize, here in the wilderness, we and they are equals.

A social creature, the wolf is not really the ruthless terror of fairy tale. But try to tell that to a snowshoe hare or other small mammal that knows its future is shortlived once it has been scented by a hungry wolf in winter.

Grey Wolf   *Crossing the Ridge*
36" x 24" (91 cm x 61 cm)

Wherever you find small fish, you'll find Great Blue Herons, whether in the wetlands of the Rockies, the coastal shoreline or in your backyard goldfish pond. These elegant birds, long and stately, move with measured grace or stand patiently staring into the water surrounding them. Their long and supple lines are an artistic inspiration.

Great Blue Heron   *Early Spring* (detail)
18" x 24" (45 cm x 61 cm)

This painting tells a story, one played out often in the wild by the predator and the prey. The wolves are following their keen sense of smell by travelling into the prevailing wind on the crusty top of the snow at the edge of the forest. Like the deer in the painting, I watched, motionless, totally captivated as they passed in silence.

A white-tailed deer relies upon its tan-colored hide to camouflage it amongst the tree trunks. The shy white-tail is well adapted to survival in the Rocky woodlands. It has keen eye-sight, hearing and sense of smell attuned to the hazards of the forest. There is plenty of the low-level vege-tation on which it forages, and its speed, agility and watchfulness make it an elusive prey.

White-tailed Deer   *Whitetail and Wolves*
24" x 36" (61 cm x 91 cm)

It is December in the Rockies and another winter storm roars past the cliffs and through the passes. This is no place for a human, and even the surefooted mountain goats stand together, waiting for the storm to dissipate. Their long-haired winter coats, covering a warm inner layer of fleece, shake and blow with the gusts of wind. Stll, they move about with incredible sureness. Flying snow stings their eyes. The wind buffets their sides and screeches past their heads as it tries unsuccessfully to push them from their precarous foothold.

Not really a goat but an antelope, this agile creature is often spied on a seemingly impossible perch high up on near-vertical cliffs. There it feeds on mosses and other sparse vegetation. The hazards of its habitat do mean that it is usually left in peace, safe from all but the most adventurous predators.

Mountain Goats   *Winter's Fury*
22" x 36" (56 CM x 91 CM)

The elusive lynx is rarely seen, preferring to prowl through the forest at night. That is a good thing, too. This compact, muscular, stub-tailed, tufted-eared keg of dynamite is not something you would like to stumble into in the dark. The lynx may be smaller than the cougar, but it can be just as ferocious, fully capable of pulling down an unlucky deer. For the most, though, it is the nemesis of the snowshoe hare. The populations of hare and lynx ebb and flow, plentiful and sparse in relation to each other, as they must have done for millennia.

Lynx   *Silent Forest*
14" x 28" (40 cm x 80 cm)

Selected Readings

Anderson, T. *Black Bear: Seasons in the Wild.* Voyageur Press.

Bauer, E. *Bears: Behavior, Ecology, Conservation.* Raincoast Books.

Bauer, E. *Whitetails: Behavior, Ecology, Conservation.* Voyageur Press.

Bauer, E. and P. *Elk: Behavior, Ecology, Conservation.* Voyageur Press.

Bauer, E. and P. *Wild & Free: The Great Wild Animals of North America.* Raincoast Books.

Berger, T. R., ed. *Majestic Elk.* Voyageur Press.

Cox, D. *Elk.* Chronicle Books.

Dennis, R. *Loons.* Voyageur Press.

Dregni, M., ed. *Loons: Song of the Wild.* Raincoast Books.

Gibson, N. *Wolves.* Raincoast Books.

Gordon, D. *Field Guide to the Bald Eagle.* Sasquatch Books.

Grambo, R. L. *Mountain Lion.* Raincoast Books.

Link, M., and K. Crowley. *Following the Pack: The World of Wolf Research.* Voyageur Press.

Majestic Whitetails. Voyageur Press.

Mech, L. D. *Way of the Wolf.* Voyageur Press.

Shirahata, S., and P. Morrow. *The Rocky Mountains.* Raincoast Books.

Snyder, N., and H. *Raptors: North American Birds of Prey.* Raincoast Books.